Praise for THREE ROADS TO QUANTUM GRAVITY

"Lee [Smolin] is a brilliant, original thinker."
　　—Roger Penrose, quoted in *New York Times Magazine*

"It would be hard to imagine a better subject."
　　—*Scientific American*

"[Smolin] argues lucidly and effect should pause to reevaluate exactly what they mean when they use the words 'space' and 'time.' This is a deeply philosophical work."
　　—*New York Times Book Review*

"Provocative, original, and unsettling."
　　—*New York Review of Books*

"More than just another work of popular science: it is a serious attempt at clarifying the author's own thoughts about the significance and interpretation of the highly mathematical theories he is discussing. The book belongs to a new genre of science writing, in which the author also tells the story of his own involvement in the research, so giving it a striking freshness."
　　—*London Review of Books*

"A mix of science, philosophy, and science fiction, [this] is at once entertaining, thought-provoking, fabulously ambitious, and fabulously speculative."
　　—*New York Times*

"An excellent writer, a creative thinker."
　　—*Nature*

"He does an excellent job of conveying the scope of current frontiers of knowledge and the excitement they have engendered. Using clear, accessible language, he takes the reader on a mind-boggling tour of three possible roads to a final theory. . . . Absolutely compelling."
　　—*American Scientist*

"This is real twenty-first-century science."
　　—Allen Lane, *Independent*